PRACTICAL GUIDE
FOR SOLAR
INSTALLATIONS

Photo voltaic and electrical wiring manual

Hallie F. Morgan

Table of Contents

CHAPTER ONE

INTRODUCTION TO SOLAR

Solar is a device that is use to draw or get light which is use in form of electricity. Today our existing day world wishes electrical energy for pretty a variety day to day functions such as industrial manufacturing, heating, transport, agricultural, lightning applications, etc.

Most of our energy desire is usually at ease via non-renewable sources of electrical energy such as coal, crude oil, natural gas, etc. But the utilization of such belongings has induced a heavy have an effect on our environment. Also, this shape of electricity beneficial useful resource is no longer uniformly disbursed on the earth.

Increase demand of renewable

There is an uncertainty of market costs such as in the case of crude oil as it depends upon on manufacturing and extraction from its reserves. Due to the restrained availability of non-renewable sources, the demand for renewable sources has grown in contemporary years. Solar power has been at the core of activity when it comes to renewable electricity sources.

Photo voltaic strength capacity

The picture voltaic standalone PV laptop as verified in fig 1 is one of the strategies when it comes to quality our electricity demand independent of the utility. Hence in the following, we will see briefly the planning, designing, and set up of a standalone PV system for electrical strength generation.

CHAPTER TWO

PHOTO VOLTAIC MECHANISMS USAGE

Steps In Installing Photo Voltaic Solar

Select location where sun reflex easily

It ought to be made wonderful that the chosen internet site each at rooftop or flooring ought to now not have hues or want to no longer have any form that intercepts the image voltaic radiation falling on the panels to be installed. Also, make nice that there won't be any structural improvement shortly surrounding the set up that can also intent the problem of shading. The flooring place of the internet site on line at which the PV set up is intended ought to be known, to have an estimation of the dimension and volume of panels required to generate the required energy output for the load. This moreover helps to plan the set up of inverter, converts, and battery banks.

Place at rooftop and perpendicular

In the case of the rooftop set up the form of roof and its structure need to be known. In the case of tilt roofs, the point of view of tilt have to be recounted and crucial mounting have to be used to make the panels have greater incidents of picture voltaic radiation i.e. ideally the radiation mindset want to be perpendicular to the PV panel and certainly as shut as to ninety degrees.

Link connection of cables and inverters

Possible routes for the cables from an inverter, battery bank, fee controller, and PV array have to be deliberate in a way that would have minimal utilization of cables and minimize voltage drop in cables. The trend dressmaker ought to select out between them and the price of the system.

Calculate the hours load in sunlight

Kilowatt-hours per rectangular meter per day KWh/m2/day: It is an extent of electricity measured in kilowatt-hours, falling on rectangular meter per day. Number of hours in a day, at some point of which irradiance averages W/m2, Peak photo voltaic hours are most in many situations used as they simplify the calculations. Do now not get harassed with the Mean Sunshine Hours and Peak

Sun Hours which you would accumulate from the meteorological station.

Insulate charger to panel

The Mean sunshine hours indicates vary of hours the sunshine's had been as the Peak solar hours is the real volume of electrical energy acquired in KWh/m2/day. Amongst all months over a size of 12 months use the lowest recommend every day insulation charge as it will make advantageous that the computing device will feature in a larger reliable way when the photo voltaic is least due to unsuitable local weather conditions.

Power rate and analyze the output

For deciding on the suitable inverter every core and output voltage and current rating has to be specified.

The inverter's output voltage is exceptional by using way of the system load; it need to be in a function to cope with the load cutting-edge and the cutting-edge taken from the battery bank. Based on the entire associated load to the device the inverter power rating can be specified.

Analyze the percentage

The price controller rating has to be 125% of the photovoltaic panel rapid circuit current. In distinct words, it has to be 25% elevated than the speedy circuit contemporary of picture voltaic panel. The size of Inverter ought to be 25% large than the whole load due to losses and hassle in the inverter. In exceptional words, it ought to be rated 135% than the entire load required in watts.

The System Voltage descriptions

The inverter enter voltage is referred to as the compute device voltage. It is moreover the regular battery pack voltage. This computer voltage is decided via way of the chosen personality battery voltage, line current, most allowable voltage drop, and strength loss in the cable. Usually, the voltage of the batteries is 12 V so will be the computer voltage

The use of Automatic Inverter UPS

Power failure and emergency breakdown can additionally take vicinity any time due to rapid circuit, damage to electric powered transmission lines, substations or specific elements of the distribution gadget, storms and exclusive lousy local weather conditions etc. In this case, emergency generator or battery backup can be used to repair the electric powered power to the domestic and exclusive associated appliances.

Work Load Performance Of Automatic Inverter UPS

In some case, it is very vital to repair the strength as rapidly as workable like in Hospital ICU, military, Intelligence and security buildings and places of work etc. This is the vicinity we use the generator and Inverter UPS

Uninterruptible Power Supply gadget with the help of backup batteries and inverter.

Steps In Installing UPS Inverter

Disconnect active line

To be part of an inverter UPS to the home electric powered supply system, First of all, disconnect these Live Line wires of two circuit breakers from the imperative distribution board which are linked to the most important double pole swap of these particular rooms which you want to be a part of to the computerized supply in each situations from battery and necessary utility electricity besides any interruption. Suppose, you desire to be part of totally two rooms and their load with UPS computerized laptop as validated in.

Connect to the board

You will have to disconnect remain wires of these rooms from the vital electrical energy grant distribution board. Now be a part of these two continue to be wires of this specific room which have to be linked to the UPS System to the output of UPS by using the two single pole MCBs separated from the important panel board.

Keep in notion that entirely the two associated MCBs and their related and linked load inverter will furnish continues energy in case of blackout.

Link the panel

To value the Battery with the resource of inverter, be a part of the Inverter UPS to the outgoing of necessary double pole (DP) MCB through a three Pin Power Plug and three Pin Power socket to the foremost supply. To be in blanketed mode, use 6 AWG 7/064" or 16mm2 cables and wire dimension to be part of the UPS to the important panel board.

Furnish the circuit in UPS

The circuit completely two rooms of the home are depends upon on the UPS and Batteries as properly as main furnish to preserve the uninterruptible electricity to the associated domestic gear and load such as lights fixtures elements and followers and so on and the distinctive loads are fed up with the resource of utility power only. Once you get the essential thinking of UPS connection, proceed to apprehend how it works in every cases i.e.

operation of the circuit when utility electrical energy is handy and battery backup as secondary power in case of power failure.

CHAPTER THREE

ROLL UP DEVICES

How to install meter 230V, 1-Phase

Use single core

Double insulated for all cables. 900mm clearance residence in width need to be furnish in case of electric powered meters and Low voltage panels and switchgears with a rate of one hundred Amps. Use minimal of 4mm2 stranded copper wire between the consumer unit and electric powered meter.

Insulate breaker

The meter tails cables used to be a part of your meter to the minimize out o or main breaker have to be double insulated in dimension of 25mm and correct terminated in to the meter slots.

Hook up on exterior

The meter tails have to be the shortest possible, or at least, the measurement of meter tails from the cut-out through the metering equipment to the client unit have to continuously be no greater than three meters of cable. The meter container has to be hooked up on the exterior component of the wall which can be easily on hand from the front and no longer going to be damaged. There want to know no longer be a gas meter, telephone and distinctive utility equipment above or underneath the electric powered meter box. The meter cabinet floor hooked up or flush mount.

Steps in Single Phase wiring Board

Connect neutral and live wire

The predominant electrical wiring of bulb, followers and many others Sub-circuits and closing sub circuits in our preceding posts, so comply with the steps under to do the same as factor out, To do single phase wring electric powered distribution machine in a multi-storey building, First of all, be a part of the single-phase strength meter to the mains furnish Connect the Neutral and Live Wire from utility pole in the first two enter slots of single part energy meter respectively.

Steps in Wiring Single Phase kWh meter

Connect the MCCB Mounded Case circuit Breaker as a two pole predominant swaps to the incoming Phase or Live and Neutral L and N from 1-phase power meter. Check the underneath section for appropriate wiring color codes in accordance to your region.

Circuit breaker load

Now be a part of the outgoing Phase and Neutral L and N from MCCB Mounded Case circuit Breaker to the DP Double Pole MCB, RCD, SP Single Pole MCBs and load of each storey.

Check lines and trace

Now be a part of the RCD from DP with Phase Line and Related Neutral Link. The Outgoing Phase traces ought to be associated to the closing circuits and closing sub circuits. The equal can be carried out for Neutral Wires. Finally, join the electrical devices and domestic tools with the Earth hyperlink terminal which leads to earth electrode in the earthing and grounding device.

Distribution board and trip

Do the same steps for all three distribution boards in unique storey. Use separate foremost breaker for each and every storey and then distribute wiring in greater three or four most essential circuits in one of a type sections. besides prolong be part of the general load of one-of-a-kind story's to the essential breaker however of separate primary board, all the linked load bulbs, followers and many others would disconnect in the entire developing at as quickly as in case of tripping any circuit due to some purpose quick circuit and many others which leads to day time out the predominant switch.

Cut edge and rate voltage

In case of single circuit alternatively of three circuits, if the load of a single storey increases, the higher cutting-edge will have to go with the flow and cables would possibly additionally be overheated. Also, voltage drop would be prolonged so the rated voltage can additionally no longer gain to the electrical appliances. In that case, the linked slight bulbs would be dimmer and one-of-a-kind domestic tools may additionally no longer work properly.

Load current

Moreover, there is a risk of electric powered furnace and hazard in the developing due to overheated cables. You have to use the important breaker with relevant rating i.e. which increase the whole load current day of all distribution boards of each and every storey and use the breaker in sub distribution board in accordance to the load of that preferred storey i.e. fans, lights and so forth in single storey and so on for different storeys.

Installing of Single Phase kWh Meter

Disconnect and locate phase line

First of all, make positive to disconnect the major energy formerly than working on electrical installations. Starting from the left essential is part of incoming from transformer Phase Line wire to the 1st slot on the meter Main, Connect the incoming Neutral (N) wire to the 2nd slot on the meter. On the Load side, is part of the Outgoing Neutral in the third slot, now be a part of the outgoing Phase wire in the 4th slot and be part of it to the main distribution board.

Core line

Make positive the characteristic of meter want to be vertical on its core line. Securely tight the bolts, washers and nuts and so forth and after connecting the wires to the meter, shut the safety windows.

Flip and complete

This way, the set up work of single area meter container is effectively completed. You can additionally flip on the most important power to take seem at and verify if the complete lot is first-rate and working accordingly.

PIN = Incoming Phase or Live from the grant of provide voltage

POUT = Outgoing Phase or Line to the home predominant distribution board.

NIN = Incoming Neutral from the grant of grant voltage.

www.ingramcontent.com/pod-product-compliance
Lightning Source LLC
Chambersburg PA
CBHW070801220526
45467CB00017B/807